THE PHYSICAL BASIS OF TIRED LIGHT

Revised Edition

Albert W. McKinney III
2017 April 5

Copyright © 2017 by Albert W. McKinney III

Library of Congress Cataloging-in-Publication Data
McKinney, Albert William III (1929-)
    The Physical Basis of Tired Light, Revised Edition
    CreateSpace Independent Publishing Platform,
        North Charleston, South Carolina
    ISBN: 978-1544804347

DEDICATION

This book is dedicated to Phyllis McKinney, my wife of 66 years.

ACKNOWLEDGEMENT

This book has benefitted from a review by Dr. Jeremy Dunning-Davies, whose eagle eye spotted some unfortunate typographical errors. The author is deeply grateful for this review!

TABLE OF CONTENTS

Page	Topic
vii	PREFACE
x	Feedback
1	PROLOGUE: MYTHS IN PHYSICS
1	Two Current Myths in Physics
1	Myth 1: Communications
1	Why is this a myth?
2	Myth 2: Extrapolations
3	Why is this a myth?
3	A Better Myth for Physics
3	Why is this a myth?
4	Why is this myth of interest?
7	CHAPTER 1: PHOTONS
9	CHAPTER 2: CONNECTION THEORY
9	Particles and Radiation
9	Particles
9	Two Properties of Primitive Particles
12	The Future
12	What Does a Probe Consist of?
12	Random Events
13	Particle Interactions Involving Probes
13	Details of Particle Interactions
14	Magnetism
15	Probe-Probe Interactions
16	Particle Interactions Involving Direct Contact
16	Neutrons
17	Elementary Particles
18	Decay of Elementary Particles
19	CHAPTER 3: THE SPEED OF LIGHT

TABLE OF CONTENTS

21   CHAPTER 4: TIRED LIGHT
22   Procedure

25   CHAPTER 5: DATA

31   CHAPTER 6: AN ANALYSIS OF POINTS 71-84

33   CHAPTER 7: MEASURED MASS?

FIGURES

11   Figure 1: Successive Probe Directions
15   Figure 2: Impulses Resulting from a Probe-Core Connection

TABLES

28   Table 1
32   Table 2

PREFACE

Over most of the last century, astronomers have collected an amazing amount of information about the stars and galaxies that surround us. This information has proved troublesome, in that it has led to such weird concepts as an expanding universe and the Big Bang. It has also led to the conclusion that the stars in remote galaxies move so fast that gravity cannot hold them together, hence the need for dark matter and/or dark energy.

Is our own galaxy, the Milky Way, unique in the universe, in the sense that it does not require dark matter or dark energy to hold it together? That seems absurd. Yet is it reasonable to question the accuracy of such a huge amount of astronomical data?

Yes! And it is easy to point to the culprit! It is the redshift z. The main reason that remote stellar velocities seem high is that their redshift values are too high.

Many years ago, Fritz Zwicky pointed out that it would seem that light lost energy as it passed through space. This idea was later termed *tired light*. But Zwicky was unable to explain why that should happen.

Now a reason has been found that justifies Zwicky's idea.

To explain this idea, it is necessary to consider several other concepts that lead up to this reason. The first such concept is the *photon*.

In 1905, Einstein suggested that electromagnetic waves must travel in packets, which he called light quanta. In 1916, such quanta were renamed photons by Leonard Troland, and the name has stuck.

What is a photon? Presumably it is a bundle of pure energy with zero rest mass.

But this is absurd, for several reasons. First, by Einstein's famous equation $E = mc^2$, energy and mass are inextricably related. So

PREFACE

energy without mass would seem impossible. Second, energy is not a substance, it is a *property* of a substance. And it seems ridiculous to think of a particle which consists solely of properties. An energy photon is as unreasonable as a length photon, or thickness photon. And third, energy is associated with the *movement* of mass. With no mass, what is moving?

So I assert that photons do not exist.

Then how is mass/energy transferred from one particle to another?

To find the answer, perhaps an unlikely way to start is with an electron. It has mass, in fact its measured mass is $m_e = 9.109 \times 10^{-31}$ kilograms. Its energy E_e can be calculated using Einstein's equation $E_e = m_e c^2$. It also has a frequency v_e, given by the equation $E_e = h v_e$, where h is Planck's constant; thus $v_e = 1.2356 \times 10^{20}$ cycles per second. But what is happening in a cycle?

It is common to envision an electron as a lump of matter. But a lump does not cycle. Hence that is not a reasonable form for an electron.

The best answer requires a gigantic leap of faith! What cycles is the *mass* of the electron! Think of the mass as a long string of matter which can either be rolled up in a ball or strung out for a long way. Think of an electron as a marvelous perpetual motion machine which treats its mass as if it were a yoyo. On each cycle, the mass is unrolled in some direction until it is all stretched out, and then it is rolled up again. On the next cycle, it is unrolled in a different direction, one which is almost opposite to the previous direction.

It is convenient to call the unrolling of the mass a *probe*, and to call the electron a *core*. Thus at the beginning of a cycle, the probe is emitted from the core, goes out a long way, finally stops when it reaches the end of the mass-string, then retracts to the core.

On its outward journey, the probe tip may hit (that is, come very, very close to) another core, or another probe tip. In such a case,

PREFACE

a very brief connection is formed between the probe tip and the other core, or the other probe tip. This connection allows an exchange of energy between the probe and its target.

In such a hit, the energy brought into the exchange by the probe tip equals that of the amount of "unwrapped" mass in the tip at that moment. To quantify this, let M be the maximum distance that the unwrapped probe tip can go, and let D be the distance of the tip from the core. Then the assertion is that the energy brought into the exchange by the probe tip is $E = E_e(1 - D/M)$. A consequence is that at the outer end of its journey, the probe tip has zero energy left, and thus cannot engage in a transfer of energy.

Now return to the idea of why redshift values are high. The idea behind using the redshift is the Doppler shift. That is, the spectrum of a moving object is changed, and from the change, the velocity of the object can be calculated.

Astronomical measurements of remote stars or galaxies ("clusters") generally make use of certain spectral lines. For this discussion, we assume that a single spectral line has been selected. Associated with this line are four wavelengths. These wavelengths and the symbols that will be used to represent them in this book are as follows:

λ_{std}: The wavelength as measured in the laboratory, or as measured by an observer located uncomfortably close to the remote cluster which is not moving with respect to that observer.

λ_{dop}: The wavelength as measured by an observer located uncomfortably close to the remote cluster which is moving with respect to that observer (includes a Doppler shift).

λ_{obs}: The wavelength actually observed on Earth (includes a Doppler shift and a decrease due to the distance travelled from the remote cluster).

PREFACE

λ_{est}: The observed wavelength λ_{obs} adjusted to estimate λ_{dop}.

The objective in astronomy is to use the observed wavelength λ_{obs} of a remote cluster to estimate the wavelength λ_{dop}, so that the latter can be used with the standard value of that wavelength λ_{std} to estimate the redshift z, which allows the calculation of the velocity of that cluster relative to Earth. The redshift z is calculated using the formula $z = (\lambda_{dop} - \lambda_{std})/\lambda_{std}$.

However, the observed wavelength λ_{obs} differs from the wavelength at the remote cluster, λ_{dop}, due to the decrease in energy because of distance. That is, the energy of the wavelength diminishes with the distance from the cluster, so that $\lambda_{obs} = \lambda_{dop}(1 - D/M)^{-1}$. Consequently, $\lambda_{est} = \lambda_{obs}(1 - D/M)$ is an estimate of λ_{dop}. (Why is this an estimate, and not exact? Because the value of M is itself merely an estimate, although a very good one.)

Therefore, replacing λ_{obs} by λ_{est} in the redshift equation yields a very good approximation to the redshift formula: $z = (\lambda_{est} - \lambda_{std})/\lambda_{std}$.

The following pages elaborate on these ideas.

Feedback

If you, the reader, have any comments you wish to make, or if you come across any errors (typographical, formatting, or in the theory itself), please let me know via e-mail at mickeymck@prodigy.net. My hope is to respond to all such messages, but I warn you that my wife and I sometimes take long trips. Hence please do not despair if a response is not immediately forthcoming!

PROLOGUE: MYTHS IN PHYSICS

Underlying any scientific theory is a set of ideas which cannot be proved or measured. They are the postulates on which the theory is based. Belief in these ideas is necessary in order to accept the theory. For better or worse, such belief is simply a matter of faith, due to the lack of any kind of proof or measurement.

According to the Wiktionary:

http://en.wiktionary.org/wiki/Wiktionary:Main_Page,

one definition of *myth* is "A person or thing existing only in imagination, or whose actual existence is not verifiable."

Thus it seems fair to say that these ideas or postulates comprise a myth.

Two Current Myths in Physics

Myth 1: Communications

One myth on which modern physics is based holds that communication between particles is accomplished either by extremely small particles, such as photons, gravitons, neutrinos, antineutrinos, quarks, and gluons, or by fields. This myth is a foundation of the standard model of modern physics.

Why is this a myth?

There are several reasons. Consider the photon. It travels from its source to its target at the speed of light. Between those two points, it is invisible. There is no way to detect its passage except by intercepting it. But intercepting it causes it to deposit its energy and then cease to exist. If it is not intercepted, then when it reaches its target, it deposits its energy and then ceases to exist.

Does the photon have mass? Apparently not, yet it has energy. How can a particle not have mass? What does it mean to say that a particle consists of pure energy?

Special relativity claims that the amount of time it takes for the photon to travel between its source and its target is zero in its own time frame. Yet to an observer, it takes d/c seconds, where c is the speed of light, and d is the distance between the source and the target. It is not obvious that these different times make sense.

But taking this a step further, it seems to say that in its own time frame, the photon exists for exactly zero seconds. Again, it seems incredible to think that a particle can consist of pure energy and only exist for zero seconds.

The same considerations apply to gravitons and other similar particles.

The above ideas imply that there is no physical evidence proving that photons, gravitons, etc., exist in the form of particles. And of course, there is no physical evidence to the contrary.

Myth 2: Extrapolations

It is traditional to consider that such things as forces and fields are independent entities, and to extrapolate their properties to cases of small distances.

For example, consider the force of gravity. It is well known that the gravitational force between two masses separated by a large distance varies with the inverse square of the distance. Extrapolating that to very small distances leads to the conclusion that gravitational force increases without limit as the masses approach each other.

Another force that is traditionally treated in a similar manner is electromagnetic force. Again, it is clear from experimental evidence that for two charged particles separated by a large distance, the electromagnetic force between them varies with the inverse square of the distance. Extrapolating that to the case of very small distances leads to the conclusion that electromagnetic force increases without limit as the two charged particles approach each other.

Why is this a myth?

These forces have never been measured at near zero distances. But it has been assumed that such increases apply in the formation of black holes. It is supposed that when a star reaches some critical mass, it encounters an extreme gravitational force which causes it to shrink instantaneously to a point mass.

Of course, such an event has never been observed.

A Better Myth for Physics

A different myth has been proposed as a basis for physics, one which replaces both of the above myths. This myth asserts that communication between particles takes place by means of the vibration of electrons, protons, and their antiparticles. This myth is the foundation of *connection theory*, which is basic to the aim of this volume.

Specifically, the assertion is that at the beginning of each vibration, a particle emits a probe, and that the probe can make a connection with another particle. During such a connection, the connected particles generally exchange an impulse, and sometimes exchange additional energy. Such exchanges are governed by the three laws of conservation (energy, linear momentum, angular momentum), which thus cause changes in the velocities and directions of the two particles.

Why is this a myth?

This is a myth because nobody has ever measured the way in which particles vibrate, nor is it likely that anybody ever will. (Is this implied by the Heisenberg Uncertainty Principle?)

MYTHS IN PHYSICS PROLOGUE

Why is this myth of interest?

There are many reasons. For one, it leads to the conclusion that when two particles approach each other, the gravitational force between them tends to a finite limit.

Likewise, it leads to the conclusion that when an electron and a proton approach each other, the electrical force between them tends to a finite limit.

Furthermore, it offers a simple, straightforward definition of gravity. That definition carries with it a number of situations in which gravity is blocked. One such blockage accounts for the precession of the perihelion of every planet and asteroid that orbits the Sun.

Also, that definition precludes the possibility that gravity causes space to be curved. Thus although the theory of general relativity provides a rather accurate approximation to some effects of gravity, it is not an accurate picture of what gravity really is, nor how it works.

Another reason why this myth is of interest is that it yields a unified description of three physical forces, showing that gravity, electromagnetism, and the strong nuclear force all arise from the same physical mechanism.

Perhaps the strongest reason for accepting this myth is that it shows why the decay of an excited hydrogen atom to the ground state takes many billions of electron cycles, and why the decay of para-positronium takes about 15 billion electron cycles.

In the standard model of physics, it would seem that in each of those cases, the atom waits all that time in order to release a photon or two. What is it waiting for?

I believe that the correct answer is this: It is waiting to find a suitable recipient for the photon energy. Apparently it is not able to send out a random photon to seek for one. Instead, it sends out its probe billions of times, looking for a connection which will allow it to release the energy of its photon.

In connection theory, the idea of energy being transmitted by probes suggests that a suitable recipient is one which has a special relation to the probe.

Therefore, connection theory defines two types of connections, which will be called loose and tight. A loose connection does not allow for the transmission of energy such as that of a photon. A tight connection is one which does allow such a transmission.

The key to a tight connection may lie with the form of the probe tip, which must align with the target core in a very special way.

So seeking such a special connection may well explain why it takes billions of connections in the above examples.

CHAPTER 1: PHOTONS

The universe contains *matter*. (Matter is stuff made up of electrons and protons, for the most part.) Matter has various properties. Two of these are *mass* and *energy*. By Einstein's famous formula $E = mc^2$, energy E is proportional to mass m. That is, energy and mass are *related*.

The wave packet was introduced by Einstein in 1905 as the means by which energy is transferred from one particle to another. (It was later renamed photon.) Thus, a photon was presented as a bundle of energy. But how can there be energy without mass? Energy is a property, not a substance. It makes just as much sense to conceive of a length-photon, that is, an object without mass which consists only of lengths.

So the idea of an energy photon is unreasonable.

Lacking photons, how is light (or any other form of energy) transferred from one particle to another? The only way that makes sense is by direct contact.

So how is energy transferred from one particle to another when the particles are not adjacent?

Well, consider that particles such as electrons and protons have *frequencies*. They vibrate!

What do these vibrations consist of?

It seems reasonable to consider that such a particle consists of two parts, which I call the core and the probe. The core is the center of mass of the particle, while the probe is an object which oscillates about the core. On its outward journey from the core, the probe may come in contact with another core, or with another probe. Such a contact allows for the exchange of energy between the probe and that other core or probe.

PHOTONS **CHAPTER 1**

Thus the probe carries energy from its core to another particle, and transfers it by direct contact.

Although photons do not exist, the word *photon* will be used in the remainder of this book to denote the energy transferred in a connection.

CHAPTER 2: CONNECTION THEORY

Particles and Radiation

The universe contains particles and radiation. As it happens, every known example of radiation originates in a particle, and its effect becomes apparent when it is deposited on another particle. Thus every kind of radiation amounts to the transfer of energy from one particle to another. Hence it is convenient to begin the story of the universe with a consideration of particles.

Particles

There are several different types of particles, some having properties that will be described below. Among these properties are ones that enable particles to communicate with one another. Such communications lead to interactions which are governed by the three basic conservation laws of physics, namely, the conservation of mass, linear momentum, and angular momentum.

To begin with, there are four types of particles that turn out to be the building blocks of matter. These four types are electrons, protons, and their antiparticles, and will be called *primitive particles*. All other particles are combinations of primitive particles. Details on such combinations are given in the section below on *Particle Interactions Involving Direct Contact*.

Two Properties of Primitive Particles

A primitive particle has two important properties. First, it has a *sign*: it is either *positive* or *negative*. Thus electrons and antiprotons are negative, while protons and positrons are positive. This property comes into play when particles communicate with one another. [Note: The choice of words here is arbitrary. The relevant property is really that there are two *kinds* of primitive particles. The words *positive* and *negative* are simply traditional names for these two kinds, but those two words are arbitrary; it would be just as informative to call the two kinds apples and oranges.]

CONNECTION THEORY CHAPTER 2

Second, a primitive particle *vibrates*. The number of vibrations per second is called the frequency of the particle. This is calculated from the formula $E = mc^2 = hv$, where E is the energy of the particle, m is its mass, c is the speed of light, h is Planck's constant, and v is the frequency. For example, an electron at rest has a frequency of 1.236×10^{20} vibrations per second, while a proton at rest has a frequency of 2.269×10^{23} vibrations per second. An antiparticle has the same frequency as its corresponding particle. A primitive particle in motion has a higher frequency, which is calculated by using the above formula with m replaced by $\dfrac{m}{\sqrt{1-v^2}}$, where v is the velocity of the particle.

What is the nature of these vibrations? To describe them, it is convenient to suppose that a primitive particle consists of two parts, called the *core* (or place holder or center of mass), and the *probe*.

A single vibration of a primitive particle is called a *cycle*. In each cycle, the core of the primitive particle emits a probe. The probe moves out in some direction from the core until either (1) it runs out of time, or (2) it connects with another particle, or (3) it connects with another probe. (A connection occurs when the tip of the probe passes close enough to another particle, or to the tip of another outgoing probe.) In the latter two cases, there is an exchange of energy. In cases (1) and (2), the probe then returns to its core, and another cycle begins. In case (3), the probe continues.

Probes are the primary means by which particles communicate with one another. (Another means of particle communications is direct contact; this is described in the section below on *Neutrons*.)

Probes are emitted in a plane which is perpendicular to the spin axis of the particle. Of course, an interaction between two particles often changes the directions of the two spin axes, so over the long run, probes are emitted in more or less random directions.

However, on two successive cycles, the probe from a primitive particle is emitted in roughly opposite directions. Since these directions are not exactly opposite, the probe directions gradually move around the

CHAPTER 2   CONNECTION THEORY

particle, an effect known as particle spin. Figure 1 (which is not to scale) indicates the directions of eight successive probes from a single primitive particle:

Figure 1
Successive Probe Directions

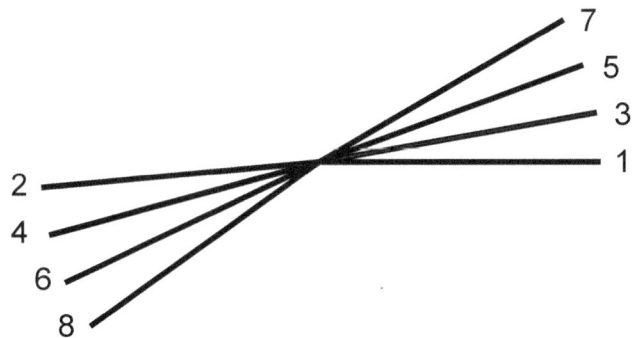

Since the cycle of an electron probe lasts for only a tiny fraction of a second, namely $1/(1.236 \times 10^{20}) = 8.09 \times 10^{-21}$ seconds, its time is up rather quickly! Even so, that is time enough for it to travel about 1.87×10^{26} meters, which is almost 20 billion light years.

The cycle of a proton probe lasts only $1/1836$ as long as that of an electron probe, and so can only go out a maximum of about 1.02×10^{23} meters, or about 11 million light years.

Since a probe obeys the rule that nothing can exceed the speed of light in a vacuum, this means that it has to move forward in time: It goes into the future! But when it returns to its parent particle, it returns to the present.

The above three paragraphs present the first three of several rather strange facts about physics!

CONNECTION THEORY CHAPTER 2

The Future

Every particle keeps its own time. When it sends out a probe, the probe goes out by its own clock, and does not go into its own future. However, its position as it travels puts it into a future relation with other particles. So there is no time travel involved here, but just a changing temporal relationship with other particles.

What Does a Probe Consist of?

Physical evidence suggests that a probe consists of the mass of the particle. It seems as if when the probe is at the core, it consists of a lump of matter with a mass equal to the mass of the particle. As it flies out from the core, it leaves behind an infinitesimal trail of mass, so that the lump which is the probe tip becomes less and less massive as the cycle continues. If a contact occurs at a distance D from the core, the reaction involves a mass $m = m_0 \left(1 - \dfrac{D}{M}\right)$ for the probe tip, where m_0 is the mass of the particle, and M is the maximum distance that the probe can go. If the probe does go that maximum distance, then at that point its effective mass is zero. As it returns to its core, it accumulates the mass that it left behind on its outward journey.

Random Events

For the cases that a probe connects with another particle or another probe, it is useful to have some names for the participants. The particle emitting the probe is called the source particle and its probe is called the source probe. A particle connected to the source probe is called a target particle; a probe connected to the source probe is called a target probe.

All interactions between particles are random events. For a connection to occur, the direction of the source probe must bring the probe tip to within some critical distance of the target core, or the tip of the target probe.

The probability of a probe-core connection depends on the distance D between the two particles, and on the area A of the source

CHAPTER 2　　　　　　　　　　CONNECTION THEORY

probe tip. The probability is equal to some constant κ times A, divided by the surface area $4\pi D^2$ of a sphere of radius D. Thus, the probability is $\kappa A/4\pi D^2$. This is the reason for the inverse square property of many physical forces.

The probability of a probe-probe connection is more complicated, but essentially equals the probability that both probes pass very near a given point in space at the same time.

There are three possible connections between particles. Two of them were discussed in the previous two paragraphs: probe-core connections, which are the cause of electromagnetic effects, and probe-probe connections, which are the sole cause of gravity. The third is a core-core connection, which is responsible for the strong nuclear force.

Probe-core connections and core-core connections result either in attractive or repulsive impulses, depending on the signs of the two particles. But probe-probe connections always result in attractive impulses, witness the observed bending of light as it passes the Sun.

Particle Interactions Involving Probes

Details of Particle Interactions

When a source probe connects with a target core, two things may happen: (1) the two particles *always* exchange energy; and (2) one particle *may* transfer additional energy to the other. The results of this interaction are governed by the laws of conservation of mass, linear momentum, and angular momentum. The energy exchange, and the energy transfer, occur in the form of impulses. The connection lasts only long enough for the interaction to take place. Following this, the source probe cannot engage in any more connections on that cycle, and soon returns to the source core.

Consider two particles which happen to interact in this way. If the two particles are stationary with respect to each other, then the resulting energy exchange involves two opposite impulses. These impulses are either both repulsive or both attractive, depending on whether the two particles have the same sign or not. If they have the same sign, then the resulting impulses tend to push the particles apart, whereas if they have opposite signs, then the impulses tend to pull them together. (Why are these two impulses opposite? Simply because one acts on one particle, and the other acts on the other particle, and by Newton's laws, these actions must be equal in force but opposite in direction.)

Note that the amount of energy exchanged has nothing to do with the concept of *charge*; it is solely determined by the above conservation laws.

[An implication of the above paragraph is that neither electrons nor protons nor their antiparticles have any charge—rather, they are electrically neutral! And this calls into question whether or not quarks exist, since they are presumed to have charges of $\pm \frac{1}{3}e$ or $\pm \frac{2}{3}e$. Clearly if a proton is electrically neutral, it cannot be composed of positively charged subparticles.]

Also, the amount of energy exchanged has nothing to do with the distance between the particles. This would seem to contradict common experience, which is that the farther apart the two particles are, the less effect the connection has. But the reason for less effect is only that the farther apart the two particles are, the less probable it is that there is any connection at all.

Magnetism

In a probe-core connection, if the target core is moving with respect to the source core, then an additional phenomenon arises which causes four, not two, impulses to be involved.

The first two impulses are equal and opposite ones which act along the line between the two cores, just as in the case of a stationary target. These are called *electrical* impulses. The third and fourth impulses are equal and opposite ones which act along a line different from that between the two cores, and are called *magnetic* impulses. Figure 2 illustrates the situation:

Figure 2
Impulses Resulting from a Probe-Core Connection

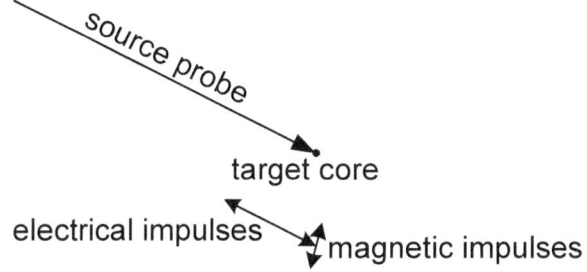

The upper and lower parts of this sketch should be superimposed, so that the intersection of the two lower arrows lies on the target core. The idea is that the electrical impulses act upon the two cores, whereas the magnetic impulses act on the target core and the source probe. The direction of the magnetic impulses depends on the direction of the target core with respect to the source core, by means of a somewhat complicated formula. Although the names differ, the electrical and magnetic impulses are physically simply impulses.

The above ideas can readily be developed to describe a magnetic field.

Probe-Probe Interactions

There are two ways in which a probe can interact with another particle. The most common is as described above, where the source probe connects with a target core.

A less common event is where the source probe connects with a target probe. In this case, the two probes exchange attractive impulses, but neither of them returns immediately to its parent particle. Instead, both continue to travel, but in slightly different directions resulting from the impulse exchange. This phenomenon is responsible for gravity, and in particular, for the gravitational bending of light passing a star.

CONNECTION THEORY CHAPTER 2

Particle Interactions Involving Direct Contact

Neutrons

The neutron is a very interesting combination of an electron and a proton. The measured mass of a neutron at rest is $m_n = 1.674\,927\,211 \times 10^{-27}$ kg, while the measured masses of an electron at rest and a proton at rest are $m_e = 9.109\,382\,91 \times 10^{-31}$ kg and $m_p = 1.672\,621\,777 \times 10^{-27}$ kg. In order for the combination to have the measured mass of a neutron, it follows that the electron must be moving very rapidly. The proton must also move rapidly, but of course it moves at about $1/1,836$ the speed of the electron, since the momenta of the two particles must be equal and opposite. Furthermore, to maintain that measured mass, neither particle can slow down.

The only possible combination is therefore one in which the electron moves in one direction at its maximum velocity $v_e = 275,367,731.239$ m s^{-1}, while the proton moves in the opposite direction at its maximum velocity $v_p = 379,328.389\,168$ m s^{-1}. These velocities continue until the next cycle begins for either particle.

When such a cycle begins, both particles exchange an impulse which, by the conservation laws, is exactly enough to cause each particle to reverse its direction but then move at the same speed as before. This impulse has the value $1.268\,946\,864\,5 \times 10^{-21}$ kg m s^{-1}.

To make this work, it seems reasonable to assume that the two particles move along a shared spin axis.

Of course, for this to work at all, it is essential that the distance between the cores of the two particles be smaller than the charge radius of a proton. Since they are so close to each other, then at the beginning of every cycle of either particle, they exchange impulses.

This is in contrast to the case where the cores are separated by more than the charge radius of a proton. In such a case, the particles do not share an impulse exchange on every cycle of either particle, but only

CHAPTER 2   CONNECTION THEORY

when the probe of either particle connects with the core of the other particle, a far less likely event.

The neutron is fundamentally unstable. The reason is that the cycle time of the electron is not an integer multiple of that of the proton. Consequently, the cycle times of the electron occur in such a way as to introduce a tiny offset between the two particles. After a huge number of cycles (which has been measured to add up to about 15 minutes), this offset builds to the point where between one cycle and the next, the two particles move apart so that they are separated by more than the charge radius of a proton.

Hence the next cycle does not cause the exchange of an impulse between the two particles. Instead, the electron finds itself traveling at extremely high velocity outside the charge radius of the proton. At its next cycle, its probe generally connects with another particle entirely, and deposits the energy as an electron antineutrino on that other particle. This has the effect of slowing the electron down to a more normal speed.

Meanwhile, the proton, now free of the electron, is also traveling at a high velocity. At its next cycle, its probe likely connects with another particle, and deposits the excess energy there, thus allowing the proton to slow to a more modest speed.

Note: Neutron decay is conventionally said to be caused by the weak interaction. But since this decay is mathematically certain, there is no force involved.

Elementary Particles

The standard model of physics seems to have two main ideas regarding elementary particles such as muons, pions, kaons, etc. First, such a particle consists of a bundle of energy (not of individual particles). Second, the bundle is held together by the strong nuclear force.

Connection theory has a different interpretation of elementary particles. First, an elementary particle consists of a close combination of primitive particles. Second, these primitive particles are held together by the strong nuclear force, constrained by the three conservation laws mentioned previously.

It has been observed that an elementary particle decays into smaller elementary particles, which then decay into yet smaller particles, ending with primitive particles. At each step in the decay, energy is released in the form of photons, neutrinos, and/or antineutrinos.

One objection to the connection-theoretic model is that it assumes that a primitive particle and its antiparticle can exist together in the elementary particle. The response to this objection is that in the most extreme case (para-positronium), which consists of an electron and a positron, it takes typically about 125 picoseconds for these two particles to annihilate each other. And many elementary particles have a half-life of a lot more than that. Hence it is reasonable to suppose that in an elementary particle with a half-life measured in microseconds, an electron and a positron could coexist quite comfortably for a short while.

Thus connection theory classifies elementary particles in terms of their decay products, which suggest the interior constitution of such a particle.

Decay of Elementary Particles

There are two different causes of elementary particle decay: *internal decay* and *escape*.

Internal decay can occur if the elementary particle contains both a particle and its antiparticle, such as an electron and a positron. Decay occurs when these two particles annihilate each other.

The constituent particles within an elementary particle A are battered about within A at each cycle of each constituent. Between every such cycle and the next, these constituents move in straight lines. Escape is where one of these particles, say P, between one cycle and the next, moves outside the charge radii of the other constituents, so that when the next such cycle occurs, the particle P is not affected by the resulting impulse, and thus escapes from A.

CHAPTER 3: THE SPEED OF LIGHT

In 1905, in the development of special relativity, Einstein decided that light was propagated by wave packets (later renamed photons). He could see no reason that a wave packet would slow down as it traveled through space. Hence he assumed that the speed of light is constant.

However, as shown in Chapter 1, the concept of a photon is unreasonable. Hence Einstein's justification for assuming that the speed of light is constant is invalid.

On the other hand, new research shows that his assumption is valid, but for a very different reason. Recall that only electrons, protons, and their antiparticles have the core-probe structure. Hence for this discussion, the word *particle* refers either to an electron, a proton, or one of their antiparticles.

Two numbers are associated with the probe of a particle: M, the maximum distance that the probe tip can travel from its core, and s, the time to complete one cycle. (Since these cycles are not periodic in the sense that the paths are repeated periodically, a cycle is defined as the time for one probe to leave the core and later return to it.) The latter number is the inverse of particle frequency v: $s = 1/v$. Thus for an electron at rest, $v_e = 1.2355 \times 10^{20}$ cycles per second, hence $s_e = 8.0939 \times 10^{-21}$ seconds. For a proton at rest, $v_p = 2.2687 \times 10^{23}$ cycles per second, hence $s_p = 4.4078 \times 10^{-24}$ seconds.

Using data compiled by Allan Sandage from 84 stars, the value of M_e has been estimated at 1.8673×10^{26} meters, which is about 19.74 billion light years. Since the frequency of a proton is about 1,836 times that of an electron, it follows that the estimate of M for a proton is about 1/1836 that of an electron, thus $M_p = 1.0170 \times 10^{23}$, or about 10.75 million light years.

THE SPEED OF LIGHT   CHAPTER 3

Thus the tip of an electron probe, if it does not encounter any probes or cores on its journey, travels out about 1.8673×10^{26} meters and back in 8.0939×10^{-21} seconds. This is about 4.6×10^{46} meters per second. Well, the universe seems to have a rule that any object which travels at a local speed greater than 299,792,458 meters per second must go on an alternate path through the universe at an apparent speed of 299,792,458 meters per second, and into the future at a rate of clock time equal to the distance traveled so far divided by 299,792,458.

A similar calculation can be made for proton probes.

Thus it is evident that probes through space are always traveling at 299,792,458 meters per second, except when they are nearly to their maximum distance, at which point their local speed drops below that value as they finally come to a stop. Thus Einstein's assumption is essentially correct.

CHAPTER 4: TIRED LIGHT

Cosmology has evolved over the centuries. Two thousand years ago, the prevailing cosmology was geocentric. It was then obvious to most people that the Moon, the Sun, and the stars revolved around Earth.

In the 16th century, Nicolaus Copernicus made the case for a heliocentric cosmology: Everything in the solar system revolved around the Sun.

Today, a further evolution in cosmology has been forced upon us. Astronomical observations have indicated that stars in galaxies other than our own are moving so fast that gravity cannot hold them together. And the farther they are from Earth, the faster those stars are moving.

To explain the reason for this, scientists have come up with the idea that there must be dark matter (or dark energy) holding them together.

Thus our own galaxy (the Milky Way) seems to be in a favored position in the universe. It is the only galaxy, or at least one of a tiny number out of the estimated two trillion galaxies in the universe, which does not seem to require any dark matter. Thus cosmology now seems bent on the idea that the universe somehow favors our galaxy, if not being centered on it.

It seems rash to think that our galaxy is so special. But what about the huge number of observations that have led to this situation? How can so many astronomers be wrong?

A simple answer to that question was suggested almost a century ago by the astronomer Fritz Zwicky. His suggestion was that photons encountered things that slowed them down as they passed through space. This idea was later termed *tired light*.

But nothing was ever found which could accomplish that.

However, with the concept of probes, another idea is suggested, namely, that as a probe goes outward, it loses energy. This can be explained in a very straightforward way: As the probe leaves its core, it leaves part of the mass of the particle behind it. Think of it as like a rubber band stretching out. The farther it goes, the more mass is "stretched" behind it. Consequently, when it engages in a contact with another probe or another core, it has only the remaining mass to use for the contact.

In less prosaic language, the assertion is that after going a distance D out of its maximum distance M, it has only the original mass m times $1-\dfrac{D}{M}$ left to use in a contact.

Thus, after traveling a distance D, the remaining energy E is then a fraction of the original energy E_0:

$$E = E_0\left(1-\frac{D}{M}\right).$$

It is expected that when this formula is used before calculating the redshift of remote stars or galaxies, the results will show that the corresponding velocities are low enough that gravity can hold remote galaxies together. Hence the need for dark matter or dark energy vanishes.

Procedure

Let E_{dop} be the energy of a probe at the beginning of a cycle on a remote star, and let E_{obs} be the energy of that probe when it reaches its target (Earth). Proposal:

$$E_{obs} = E_{dop}\left(1-\frac{D}{M}\right),$$

CHAPTER 4 — TIRED LIGHT

Hence estimate E_{dop} by reversing the above equation to get E_{est}:

$$E_{est} = E_{obs}\left(1 - \frac{D}{M}\right)^{-1}.$$

(This is an estimate because the value of M is an estimate.) Expressed in terms of wavelength, the equation is

$$\frac{ch}{\lambda_{est}} = \frac{ch}{\lambda_{obs}}\left(1 - \frac{D}{M}\right)^{-1},$$

or more appropriately

(4-1) $$\lambda_{est} = \lambda_{obs}\left(1 - \frac{D}{M}\right)$$

Then calculate the redshift z from

$$z = \frac{\lambda_{est} - \lambda_{std}}{\lambda_{std}}.$$

where λ_{std} is the expected value for λ when the remote object is at a constant distance from Earth (neither receding nor approaching).

These steps lead to a value of z which compensates for the loss of energy of the probe on its outward journey.

TIRED LIGHT CHAPTER 4

CHAPTER 5: DATA

Table 1 at the end of this chapter uses data from Tables 2, 3, and 4, of the paper by Allan Sandage in the Astrophysical Journal, V. 178, pp. 6-29. The data are arranged in order of increasing distance from Earth.

In that paper, Sandage refers to several manuscripts in which various colleagues made calculations in support of the paper. I presume that these manuscripts included calculations of the redshift values used in the paper. Such calculations would have taken the form

$$z = \frac{\lambda_{obs} - \lambda_{std}}{\lambda_{std}}$$

where z is the redshift, λ_{obs} is the observed wavelength, and λ_{std} is the standard wavelength for that spectral line.

Equation (4-1) of the previous chapter must be applied to λ_{obs} before the above redshift formula is used. But the values of λ_{obs}, presumably in those manuscripts, were not readily available to me at the time of writing this book.

In lieu of those values, I took three arbitrary wavelengths, namely 3819Å, 4144Å, and 4481Å. To estimate λ_{obs}, I solved the equation

$$\lambda_{obs} = \lambda_{std}(z+1)$$

for each of those three standard wavelengths, and for each data point in Table 1.

For each of the resulting estimated values of λ_{obs}, I applied the formula from the previous chapter to get a corrected value:

DATA CHAPTER 5

$$\lambda_{est} = \lambda_{obs}(1 - D/M),$$

where D is the distance of the cluster from Earth as given in Table 1, and M is the estimated maximum distance that light can travel. In a previous book of mine, the value of M was estimated to be 1.8673×10^{26} meters.

Each such corrected wavelength λ_{est} was then used to calculate the corrected redshift for that cluster:

$$z_{corr} = \frac{\lambda_{est} - \lambda_{std}}{\lambda_{std}}.$$

It turned out that the results were the same for each of the three wavelengths used for λ_{std}. Therefore, only the results for the middle wavelength, 4144Å, are reported in Table 1.

A point by point plot of the raw redshift values (those from Sandage's paper) is shown below, followed by a plot of the corrected redshift values.

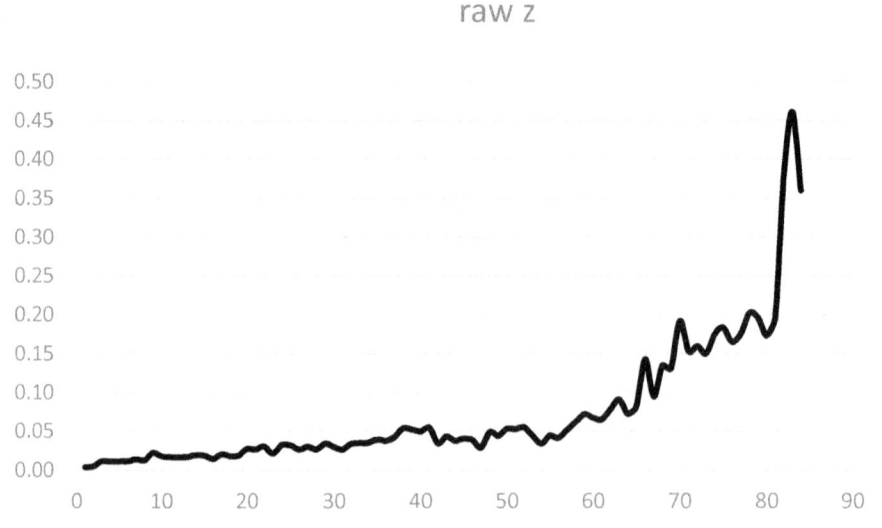

CHAPTER 5 — DATA

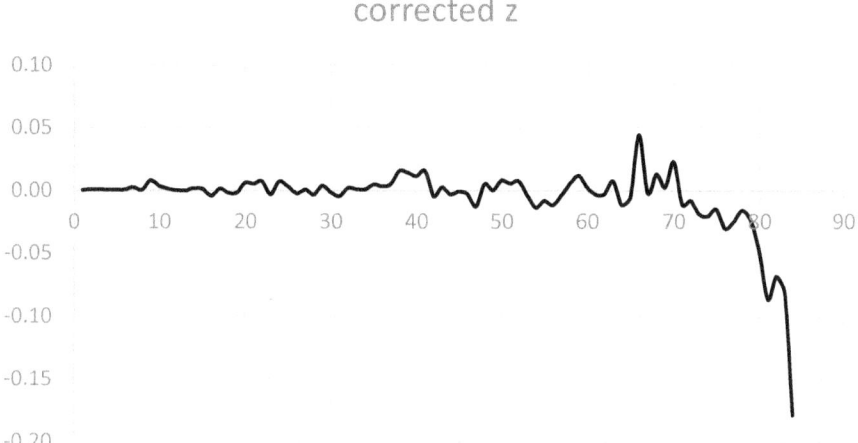

Whereas the raw z values increase almost exponentially, the corrected values wobble about 0, at least for the first 70 points. After that, the data are not sufficiently accurate to show any reasonable trend. Of course, the distances from Earth of points 71 to 84 are greater than 2.8 billion light years.

Thus Sandage's data provide a vary good fit for the formula from the previous chapter. Since that formula was derived from theoretic considerations, and without any data, I assert that the data fit proves that the formula is reasonable, and that the behavior of an electron as described in earlier chapters is also justified.

Table 1 follows. The first line contains the values of the constant M (in units of 10^{25} meters), and the arbitrary wavelength used for λ_{std}. From there on, the columns of the table consist of (1) the point numbers, (2) the redshift values from Sandage's paper, (3) the distances of the points from Earth (also in units of 10^{25} meters), derived from Sandage's paper, (4) the observed wavelength as inferred from a formula above, (5) the estimated wavelength (an estimate of λ_{dop}), applying the formula from the previous chapter, and (6) the corrected redshift values/

DATA

TABLE 1

	M	18.673	λ_{std}	4144	
n	z	D	λ_{obs}	λ_{est}	z_{corr}
1	0.00379	0.0638	4160	4145	0.00036
2	0.00509	0.0785	4165	4148	0.00086
3	0.0114	0.1945	4191	4148	0.00087
4	0.0113	0.1972	4191	4147	0.00062
5	0.0115	0.2000	4192	4147	0.00067
6	0.0118	0.2028	4193	4147	0.00081
7	0.0138	0.2075	4201	4155	0.00253
8	0.0128	0.2265	4197	4146	0.00051
9	0.0222	0.2612	4236	4177	0.00790
10	0.0181	0.2648	4219	4159	0.00366
11	0.0170	0.2851	4214	4150	0.00147
12	0.0168	0.3041	4214	4145	0.00024
13	0.0169	0.3126	4214	4143	-0.00012
14	0.0193	0.3228	4224	4151	0.00168
15	0.0188	0.3243	4222	4149	0.00111
16	0.0145	0.3443	4204	4127	-0.00421
17	0.0204	0.3491	4229	4149	0.00132
18	0.0180	0.3656	4219	4136	-0.00193
19	0.0189	0.3810	4222	4136	-0.00189
20	0.0273	0.3881	4257	4169	0.00595
21	0.0268	0.3935	4255	4165	0.00516
22	0.0303	0.4178	4270	4174	0.00725
23	0.0215	0.4498	4233	4131	-0.00311
24	0.0322	0.4539	4277	4173	0.00711
25	0.0318	0.5188	4276	4157	0.00313
26	0.0266	0.5260	4254	4134	-0.00232
27	0.0301	0.5333	4269	4147	0.00068
28	0.0267	0.5457	4255	4130	-0.00330
29	0.0341	0.5482	4285	4160	0.00374

TABLE 1
(continued)

n	z	D	λ_{obs}	λ_{est}	z_{corr}
30	0.0298	0.5662	4267	4138	-0.00143
31	0.0266	0.5662	4254	4125	-0.00453
32	0.0335	0.5714	4283	4152	0.00187
33	0.0351	0.6151	4289	4148	0.00100
34	0.0351	0.6180	4289	4147	0.00084
35	0.0391	0.6180	4306	4164	0.00471
36	0.0377	0.6180	4300	4158	0.00336
37	0.0428	0.6745	4321	4165	0.00513
38	0.0537	0.6776	4367	4208	0.01546
39	0.0522	0.6776	4360	4202	0.01402
40	0.0497	0.6807	4350	4191	0.01143
41	0.0543	0.6870	4369	4208	0.01551
42	0.0344	0.6998	4287	4126	-0.00437
43	0.0431	0.7261	4323	4155	0.00254
44	0.0376	0.7328	4300	4131	-0.00312
45	0.0404	0.7396	4311	4141	-0.00081
46	0.0387	0.7464	4304	4132	-0.00282
47	0.0287	0.7499	4263	4092	-0.01261
48	0.0487	0.7780	4346	4165	0.00501
49	0.0440	0.7816	4326	4145	0.00030
50	0.0530	0.7961	4364	4178	0.00811
51	0.0526	0.8336	4362	4167	0.00561
52	0.0549	0.8414	4372	4175	0.00737
53	0.0440	0.8531	4326	4129	-0.00370
54	0.0339	0.8531	4284	4089	-0.01334
55	0.0441	0.9354	4327	4110	-0.00820
56	0.0414	0.9484	4316	4096	-0.01149
57	0.0516	0.9840	4358	4128	-0.00382
58	0.0621	0.9931	4401	4167	0.00561
59	0.0718	1.0447	4442	4193	0.01184

TABLE 1
(concluded)

n	z	D	λ_{obs}	λ_{est}	z_{corr}
60	0.0670	1.1297	4422	4154	0.00245
61	0.0649	1.1939	4413	4131	-0.00319
62	0.0779	1.3963	4467	4133	-0.00270
63	0.0904	1.4223	4519	4174	0.00735
64	0.0722	1.4554	4443	4097	-0.01137
65	0.0825	1.5170	4486	4121	-0.00544
66	0.1426	1.6106	4735	4327	0.04405
67	0.0944	1.6405	4535	4137	-0.00175
68	0.1345	1.9998	4701	4198	0.01300
69	0.1312	2.1231	4688	4155	0.00258
70	0.1917	2.6484	4938	4238	0.02268
71	0.1532	2.6606	4779	4098	-0.01111
72	0.1594	2.6976	4805	4110	-0.00809
73	0.1499	2.7351	4765	4067	-0.01853
74	0.1745	3.0973	4867	4060	-0.02031
75	0.1831	3.1259	4903	4082	-0.01495
76	0.1651	3.1259	4828	4020	-0.02994
77	0.1745	3.1694	4867	4041	-0.02485
78	0.2018	3.3805	4980	4079	-0.01577
79	0.1959	3.4275	4956	4046	-0.02361
80	0.1703	3.5398	4861	3939	-0.04936
81	0.1979	4.4359	4964	3785	-0.08667
82	0.3800	6.0671	5719	3861	-0.06838
83	0.4610	6.9339	6054	3806	-0.08152
84	0.3600	7.3957	5636	3404	-0.17865

CHAPTER 6: AN ANALYSIS OF POINTS 71-84

Since points 71-84 did not fit into the pattern of the first 70 points, it is appropriate to see what might be the cause of this. To that end, this chapter seeks to find the distances of those cluster from Earth that would cause them to appear consistent with the first 70 points. In other words, what distances would cause their redshift values to be zero?

The analysis uses two formulas. One is Equation (4-1), which seeks to estimate λ_{dop} from the observed wavelength λ_{obs}:

$$\lambda_{est} = \lambda_{obs}\left(1 - \frac{D}{M}\right).$$

The second is the formula used to calculate the redshift z:

$$z = \frac{\lambda_{est} - \lambda_{std}}{\lambda_{std}}.$$

If the value of z is not close to zero, then it may be due to an error in the measured distance. To see what the size of the error might be, replace z by zero:

$$\frac{\lambda_{est} - \lambda_{std}}{\lambda_{std}} = 0,$$

or

$$\lambda_{est} = \lambda_{std}.$$

Now replace λ_{est} by $\lambda_{obs}\left(1 - \frac{D}{M}\right)$

$$\lambda_{obs}\left(1 - \frac{D}{M}\right) = \lambda_{std}$$

AN ANALYSIS OF POINTS 71-84 CHAPTER 6

Then solve this equation for D:

$$D = M\left(1 - \frac{\lambda_{std}}{\lambda_{obs}}\right)$$

Table 2 below shows the results. It turns out that the distance values obtained by Sandage seem much too high. In this table, the first column gives the point numbers, the second gives the distance D from Earth according to Sandage, and the third gives the distance D_0 from Earth that would lead to redshift values of zero. Of course there may be other reasons for these clusters not to conform to the behavior of the first 70 points.

TABLE 2

n	D	D_0
71	2.6606	2.4807
72	2.6976	2.5673
73	2.7351	2.4342
74	3.0973	2.7743
75	3.1259	2.8899
76	3.1259	2.6460
77	3.1694	2.7743
78	3.3805	3.1355
79	3.4275	3.0588
80	3.5398	2.7540
81	4.4359	3.0849
82	6.0671	5.1418
83	6.9339	5.8920
84	7.3957	4.9429

CHAPTER 7: MEASURED MASS?

In previous chapters, the phrase "measured mass" has been used. Here is the reason: It has been shown that the mass of an electron expands and contracts as the electron goes through its cycles. A consequence of this is that measurements of its mass vary with time. However, since the electron cycles some 10^{20} times per second, and since the measurements make use of the behavior of the electron over short intervals of time, the variability of the measurements is glossed over in the time of measurements.

A surprising conclusion is that the actual mass of the electron is twice the measured mass. The reason for this is discussed below.

If the electron does cycle its mass, then the energy of the electron varies as it goes through its cycle. Consider only the first half of a cycle. The amount m of mass that is moving varies with the distance D moved:

$$m = m_0 \left(1 - \frac{D}{M}\right),$$

where m_0 is the actual mass, and M is the maximum distance that the probe tip can move. Thus the instantaneous energy is

$$E = m_0 c^2 \left(1 - \frac{D}{M}\right)$$

Then the average energy is

MEASURED MASS? CHAPTER 5

$$ave(E) = m_0 c^2 \frac{\int_{D=0}^{M}\left(1-\frac{D}{M}\right)dD}{\int_{D=0}^{M} dD}$$

$$= m_0 c^2 \frac{D\left|\begin{matrix}M\\0\end{matrix}\right. - \frac{D^2}{2M}\left|\begin{matrix}M\\0\end{matrix}\right.}{M}$$

$$= m_0 c^2 \frac{M - \frac{M}{2}}{M}$$

$$= \frac{m_0 c^2}{2}$$

Thus it would seem that the measured mass of an electron is half of the true actual mass,

Of course, the fact that the measurements are taken in a laboratory, where the distances between the electron probe and the instrumentation are small, will likely bias the result!

www.ingramcontent.com/pod-product-compliance
Lightning Source LLC
Chambersburg PA
CBHW070136210526
45170CB00013B/1235